21st Century Junior Library

UNDERSTANDING TORNADOES

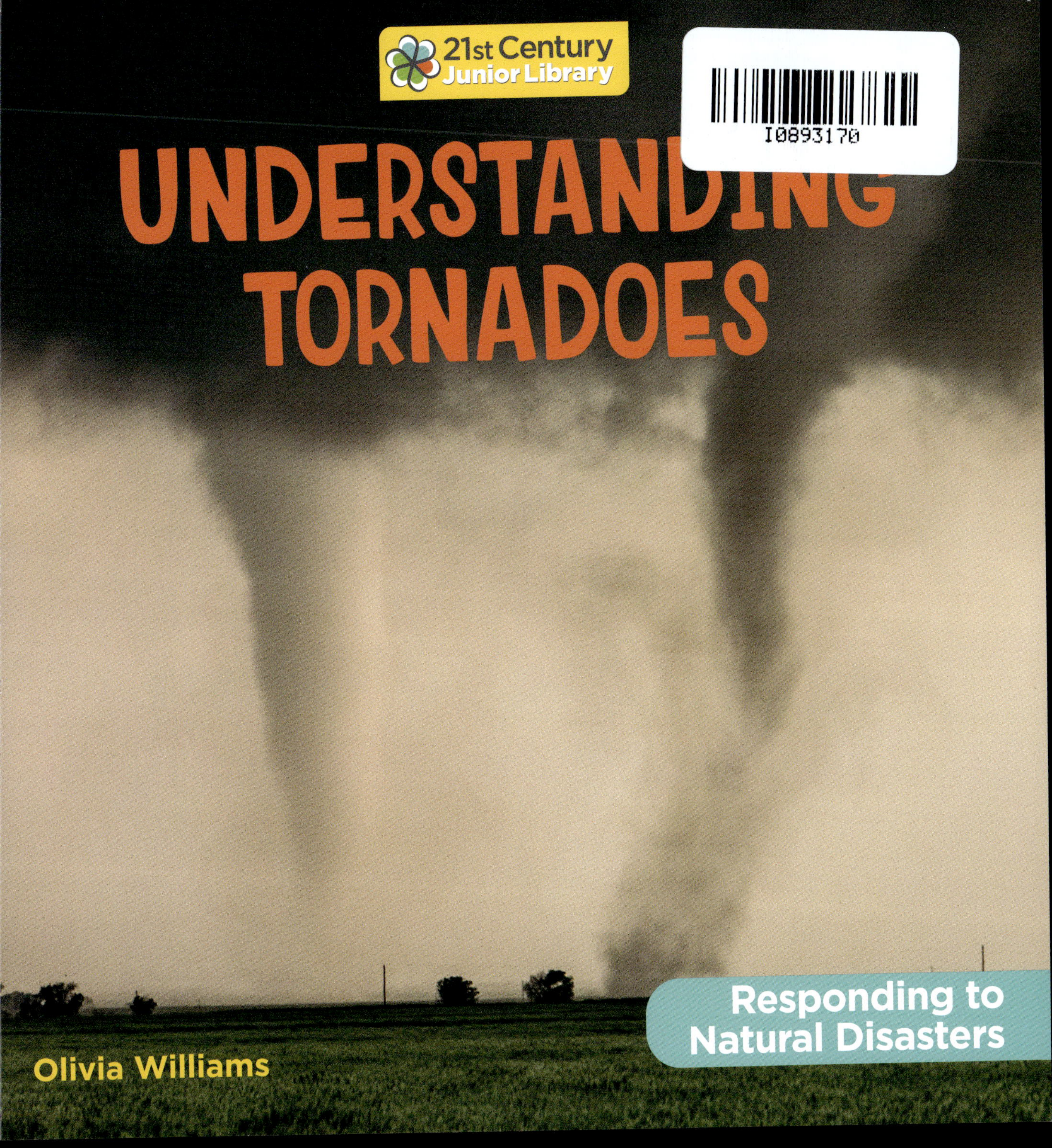

Responding to Natural Disasters

Olivia Williams

Published in the United States of America by:

CHERRY LAKE PRESS
2395 South Huron Parkway, Suite 200, Ann Arbor, Michigan 48104
www.cherrylakepress.com

Reading Adviser: Beth Walker Gambro, MS, Ed., Reading Consultant, Yorkville, IL
Photo Credits: © Eugene R Thieszen/Shutterstock.com, cover, 1; © EmiliaUngur/Shutterstock.com, 5; © Joe Belanger/Shutterstock.com, 6; © OszO/Shutterstock.com, 7; © Jeff Smith - Perspectives/Shutterstock.com, 9; © Minerva Studio/Shutterstock.com, 10, 11; © Stanislaw Tokarski/Shutterstock.com, 12 [left]; © Alexey Stiop/Shutterstock.com, 12 [right]; © GSW Photography/Shutterstock.com, 13 [top]; © Gino Santa Maria/Shutterstock.com, 13 [bottom, left]; © Newtonian/Shutterstock.com, 13 [bottom, right]; © Canterbury Stormchaser/flickr, 15 [top]; © Laura Hedien/Shutterstock.com, 15 [bottom]; © Noel V. Baebler/Shutterstock.com, 16; © Iulian Valentin/Shutterstock.com, 19

Cherry Lake Press is an imprint of Cherry Lake Publishing Group.

Library of Congress Cataloging-in-Publication Data
Names: Williams, Olivia (Writer of children's books), author.
Title: Understanding tornadoes / Olivia Williams.
Description: Ann Arbor, MI : Cherry Lake Publishing, [2022] | Series: Responding to natural disasters | Audience: Grades 2-3
Identifiers: LCCN 2021043050 | ISBN 9781534199163 (hardcover) | ISBN 9781668900307 (paperback) | ISBN 9781668901748 (pdf) | ISBN 9781668906064 (ebook)
Subjects: LCSH: Tornadoes—Juvenile literature.
Classification: LCC QC955.2 .W465 2022 | DDC 363.34/923—dc23/eng/20211005
LC record available at https://lccn.loc.gov/2021043050

Cherry Lake Press would like to acknowledge the work of the Partnership for 21st Century Learning, a Network of Battelle for Kids. Please visit http://www.battelleforkids.org/networks/p21 for more information.

Printed in the United States of America
Corporate Graphics

ABOUT THE AUTHOR

Dr. Olivia Williams is from mid-Wales in the United Kingdom. She has traveled to many countries to help people after natural disasters. She has felt earthquakes and aftershocks, and seen damage from floods, typhoons, and wildfires. The natural environment can be harsh but also beautiful. Olivia loves to hike and run in the mountains and photograph the incredible things she sees.

CHERRY LAKE PRESS

CONTENTS

WHAT ARE TORNADOES?

A tornado is a spiraling windstorm. Another popular name for a tornado is a twister. Tornadoes create the strongest winds on Earth. These winds can be as strong as 300 miles (483 kilometers) per hour.

Tornadoes normally occur during thunderstorms in the spring and summer. This is because tornadoes need cold and warm air to form. Tornadoes also

Tornadoes are also called twisters because of how they act.

need the air to be **moist**. A tornado might occur when wind blows warm, moist air into cold, dry air. A tornado can suddenly disappear when it moves over colder ground. But scientists aren't sure exactly why they die out. Scientists still need to do more research.

Tornadoes form when cold, dry air meets with warm, moist air.

Not all tornadoes can be seen, but they all can be heard!

A tornado is commonly recognized by its twisting column. One end touches the clouds. The other end touches the ground.

Tornadoes most often occur in North America. They are especially common in a region called "Tornado Alley." This region covers much of the plains area in the center of the United States. Tornadoes normally last only a few seconds to a few minutes. But some can last a few hours! The longest tornado lasted 3.5 hours in 1925. Even short-lived tornadoes can be very deadly. In April 2011, more than 300 tornadoes formed across 15 U.S. states. The lives of 348 people were lost. This event was called a "super outbreak."

Ask Questions!

Natural disasters affect people in different ways. Talk to a few family members and friends. Ask them if they've ever experienced a tornado. How did they deal with the event during and after? How do you think you'd respond to a tornado?

The United States is hit with about 1,000 tornadoes a year!

HOW TORNADOES AFFECT THE PLANET AND PEOPLE

Tornadoes are violent **forces of nature**. They can destroy houses and tear off roofs. They can throw cars into the air and flip trains. They can suck all the water from a riverbed. They can uproot trees. They can kill people and animals. Tornadoes can ruin large sections of the natural environment. Old trees can be wiped out in seconds. The homes of wild animals can be destroyed. Fields

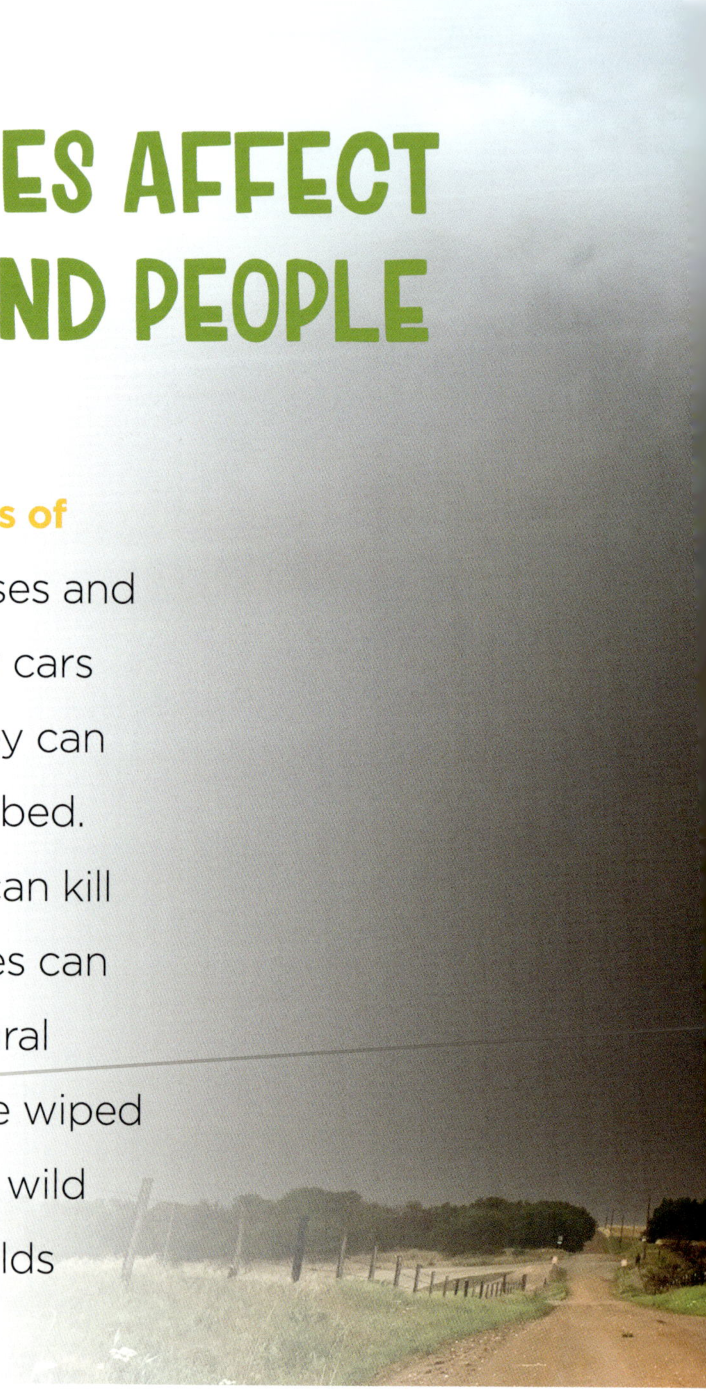

planted with crops can be ruined too. This means that food might be difficult to find after a tornado.

Tornadoes can destroy power plants and gas stations. This is dangerous because it releases **toxic** material into the **atmosphere**. Tornadoes can also destroy water treatment plants. This means that water can become unsafe to drink.

Tornadoes are unpredictable. They can change direction at any time.

The world's deadliest tornado occurred in 1989. It happened in Bangladesh in southern Asia. The storm killed 1,300 people and left 80,000 others homeless. It took a long time for people to recover because many of them were poor. They couldn't afford to buy the materials to rebuild their homes.

Create!

Tornadoes are terrifying. They can cause fear. When people are scared, it can be hard for them to remember instructions. Think of five instructions you should follow during a tornado. Create a poster with these instructions to place on your refrigerator. The poster will remind you and your family how to prepare for a tornado!

People called storm chasers like to watch and follow tornadoes. Some storm chasers get paid to chase storms. They film tornadoes and sell the footage to news stations. Even though they are experienced, they still get frightened.

Some homes have storm cellars. These are like basements but built in a way that protects people from dangerous weather, like tornadoes.

PROTECTING AGAINST AND RECOVERING FROM TORNADOES

Tornadoes can be very stressful for people who experience them. Preparation doesn't ensure that people, animals, and buildings will be safe. But it can help. The U.S. government suggests several tornado preparations. Identify a "safe room" where you can hide. The best safe room is a basement or a ground-floor room in the center of a building. Make sure the room doesn't have windows. Hide under a heavy object, like a large table. Put food and water in your safe room. Make sure you have a cell phone with you. If you're caught outside during a tornado, stay away from water. Stay away from trees and telephone poles too. Lightning is known to strike tall objects!

Hospitals rely on electricity to run medical machines. But electricity is often cut off during a tornado. People have learned from other natural disasters. This is why places such as hospitals have **generators** to produce backup electricity. Technology can help too. **Radar** systems like the pulse-Doppler radar provide lots of information

Make a Guess!

Tornadoes are often shown in movies. These movies can be difficult for people to watch. Watching natural disasters can make people feel traumatized, especially if they have experienced a tornado themselves. Why do you think that is? What kind of emotions might they show? How might they act?

During a tornado, rooms on the lowest floor and without windows, like the basement or a closet, are the safest places in a house.

about tornadoes. They tell us their size, speed, and direction. With this information, emergency responders know which towns will need aid. Airplanes also help. They have fixed devices that collect weather **data**. This data is sent to supercomputers on the ground. Because of this data, scientists understand much more about tornadoes.

Organizations and charities help after tornadoes. Doctors might provide free medical care. Food banks provide food. Families donate money to other families. Schools hold fundraisers for tornado survivors. Builders help rebuild homes. Emergency teams even fly across the world to support people after a tornado. In 2020, Tennessee experienced a big tornado. Celebrity singers held an event that raised $500,000 for survivors! When people are harmed by a natural disaster, many others want to help.

ACTIVITY

Create a Vortex

Make a "tornado" in a bottle! Discover how tornadoes work by making a mini version yourself!

Materials Needed:

- Dish soap
- Water
- Clear bottle with cap
- Glitter or food coloring (optional)

Instructions:

1. Fill the bottle more than halfway with water. Make sure to leave a bit of room at the top.
2. Add a few drops of dish soap and glitter or food coloring.
3. Seal the bottle with its cap.
4. Move the bottle in a circle. Watch the water form into a tornado!

Spinning the bottle creates a vortex. A vortex is fluid or gas (like air) that is spinning. Examples of vortexes found in nature are tornadoes, waterspouts, and hurricanes.

GLOSSARY

atmosphere (AT-muh-sfihr) the air surrounding Earth

data (DAY-tuh) information, facts, and numbers, often stored on a computer

forces of nature (FOR-suhz UHV NAY-chuhr) powerful aspects of nature and the weather, like winds and storms

generators (JEH-nuh-ray-tuhrz) machines that convert mechanical energy into electricity

moist (MOYST) slightly wet

natural disasters (NAH-chuh-ruhl dih-ZAH-stuhrz) natural events such as floods, earthquakes, or hurricanes that cause great damage

radar (RAY-dar) a device that sends out radio waves for finding out the position and speed of a moving object

toxic (TAHK-sik) poisonous

traumatized (TRAH-muh-tyzd) suffering from physical and emotional injury and distress that can last for a long time

FIND OUT MORE

Books

Challoner, Jack. *Eyewitness Hurricane & Tornado.* New York, NY: DK Publishing, 2014.

Raum, Elizabeth. *Tornado!* Mankato, MN: Amicus Ink, 2017.

Websites

Tornadoes 101—National Geographic

https://www.youtube.com/watch?v=aacHWoB7cmY
Watch and learn more about tornadoes.

Trinka and Sam and the Swirling Twirling Wind

https://www.nctsn.org/resources/trinka-and-sam-and-swirling-twirling-wind
Read about Trinka and Sam's experience of a tornado with your parents.

INDEX